Edited By: Jamia Brown-Curry

Cover Art By: Donya Curry

Layout By: Donya Curry

Photographs By: Donya Curry

Published By

DNJ CO LLC

211 Pauline Drive Ste 1039

York, PA 17402

www.dnjco.net

www.dnjcokids.com

DNJ Co Kids

DIRECTIONS: TRACE THE WORDS AND NUMBERS BELOW.

1

One

1 1 1 1 1

1 1 1 1 1 1 1 1

COLOR ONE STAR:

CIRCLE THE ONES:

1	3	5
4	2	1
6	1	8
1	7	1
2	1	4

DIRECTIONS: TRACE THE WORDS AND NUMBERS BELOW.

Two

2 2 2 2

2 2 2 2 2 2

COLOR TWO TRIANGLES:

CIRCLE THE TWOS:

2	3	5
4	2	1
6	1	2
1	2	1
2	1	4

DIRECTIONS: TRACE THE WORDS AND NUMBERS BELOW.

Three

3 3 3 3

3 3 3 3 3 3

COLOR THREE SQUARES:

CIRCLE THE THREES:

2	3	5
4	2	3
3	1	2
3	5	1
6	3	4

DIRECTIONS: TRACE THE WORDS AND NUMBERS BELOW.

Four

4 4 4 4

4 4 4 4 4 4

COLOR FOUR HEARTS:

CIRCLE THE FOURS:

1	2	3
4	5	4
3	1	2
4	3	5
2	1	4

DIRECTIONS: TRACE THE WORDS AND NUMBERS BELOW.

5

Five

5 5 5 5

5 5 5 5 5 5

COLOR FIVE CIRCLES

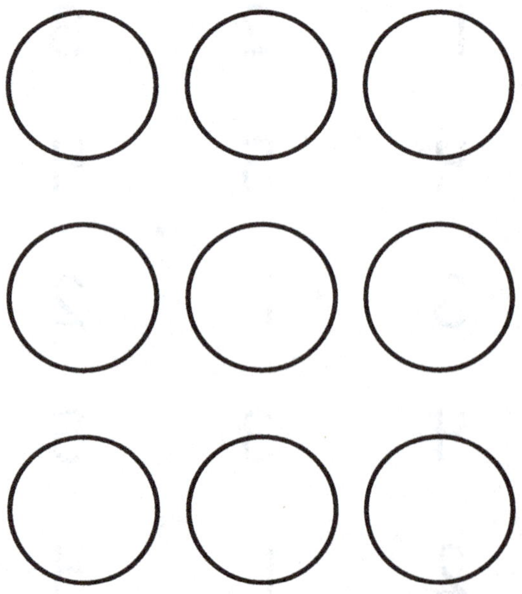

CIRCLE THE FIVES

5	2	3
1	5	4
3	5	2
4	3	5
2	5	4

DIRECTIONS: TRACE THE WORDS AND NUMBERS BELOW.

Six

6 6 6 6

6 6 6 6 6 6

COLOR SIX TRIANGLES:

CIRCLE THE SIXES

5	6	3
1	2	6
6	5	4
4	3	6
2	6	1

DIRECTIONS: TRACE THE WORDS AND NUMBERS BELOW.

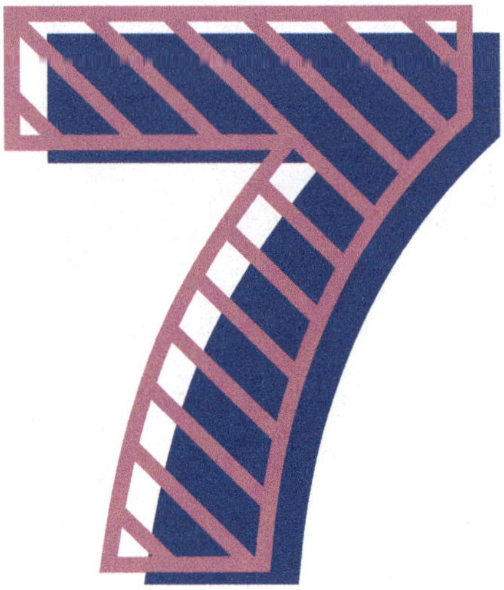

Seven

7 7 7 7

7 7 7 7 7 7

COLOR SEVEN HEXAGONS

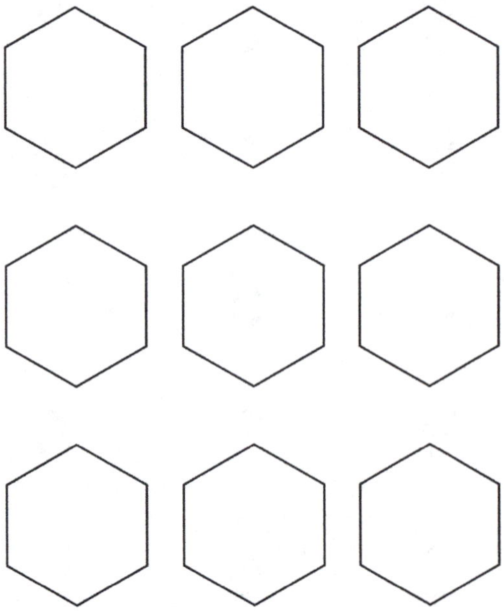

CIRCLE THE SEVENS

5	7	3
1	7	4
7	3	2
4	2	7
7	5	4

DIRECTIONS: TRACE THE WORDS AND NUMBERS BELOW.

Eight

8 8 8 8

8 8 8 8 8 8 8

COLOR EIGHT DIAMONDS:

CIRCLE THE EIGHTS:

1	8	3
4	5	8
8	6	2
4	3	8
2	5	1

DIRECTIONS: TRACE THE WORDS AND NUMBERS BELOW.

Nine

9 9 9 9

9 9 9 9 9

COLOR NINE HATS:

CIRCLE THE NINES:

9	8	6
4	7	9
8	6	9
4	9	8
9	5	1

DIRECTIONS: TRACE THE WORDS AND NUMBERS BELOW.

10

Ten

10 10 10

10 10 10 10

COLOR TEN STAR:

CIRCLE THE TENS

10	3	5
4	2	10
6	1	8
10	7	1
2	10	4

DIRECTIONS: TRACE THE WORDS AND NUMBERS BELOW.

11

Eleven

11 11 11

11 11 11 11

COLOR ELEVEN STAR:

CIRCLE THE ELEVENS

10	11	5
4	2	10
6	11	8
11	7	11
2	10	4

DIRECTIONS: TRACE THE WORDS AND NUMBERS BELOW.

12

Twelve

12 12 12

12 12 12 12

COLOR TWELVE STAR:

CIRCLE THE TWELVES

10	12	5
4	2	10
6	12	8
12	7	12
2	10	4

13

DIRECTIONS: TRACE THE WORDS AND NUMBERS BELOW.

Thirteen

13 13 13

13 13 13 13

COLOR THIRTEEN HEART

CIRCLE THE THIRTEENS

13	12	5
4	2	10
6	13	8
12	7	13
13	10	4

DIRECTIONS: TRACE THE WORDS AND NUMBERS BELOW.

14

Fourteen

14 14 14

14 14 14 14

COLOR FOURTEEN CAP:

CIRCLE THE FOURTEENS

13	14	5
4	2	14
14	13	8
12	7	13
14	10	4

DIRECTIONS: TRACE THE WORDS AND NUMBERS BELOW.

15

Fifteen

15 15 15

15 15 15 15

COLOR FIFTEEN HEXAGONS

CIRCLE THE FIFTEENS

13	14	15
4	2	14
15	13	8
12	7	13
15	10	15

DIRECTIONS: TRACE THE WORDS AND NUMBERS BELOW.

16

Sixteen

16 16 16

16 16 16 16

COLOR SIXTEEN CIRCLES

CIRCLE THE SIXTEENS

16	14	15
4	2	4
15	16	8
12	7	13
16	10	15

DIRECTIONS: TRACE THE WORDS AND NUMBERS BELOW.

17

Seventeen

17 17 17

17 17 17 17

COLOR SEVENTEEN CIRCLES

CIRCLE THE SEVENTEENS

16	14	15
17	2	17
15	16	8
12	17	13
17	10	15

DIRECTIONS: TRACE THE WORDS AND NUMBERS BELOW.

18

Eighteen

18 18 18

18 18 18 18

COLOR EIGHTEEN STARS

CIRCLE THE EIGHTEENS

18	14	15
4	17	4
15	18	17
12	7	18
18	17	15

DIRECTIONS: TRACE THE WORDS AND NUMBERS BELOW.

19

Nineteen

19 19 19

19 19 19 19

COLOR NINETEEN STARS

CIRCLE THE NINETEENS

19	14	15
4	17	19
15	19	17
19	7	18
18	19	15

DIRECTIONS: TRACE THE WORDS AND NUMBERS BELOW.

20

Twenty

20 20 20

20 20 20

COLOR TWENTY HEARTS

CIRCLE THE TWENTYS

18	20	15
4	17	20
20	18	17
12	7	18
18	20	15

DIRECTIONS: TRACE THE WORDS AND NUMBERS BELOW.

21

Twenty-one

21 21 21

21 21 21 21

COLOR TWENTY-ONE STARS

CIRCLE THE TWENTY-ONES

21	14	15
4	17	4
15	21	17
12	7	21
21	17	15

22

DIRECTIONS: TRACE THE WORDS AND NUMBERS BELOW.

Twenty-two

22 22 22

22 22 22 22

COLOR TWENTY-TWO TRIANGLES:

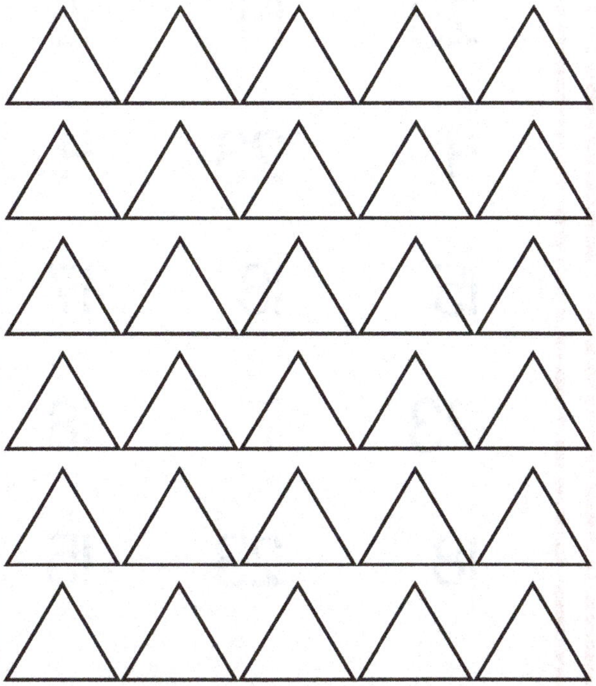

CIRCLE THE TWENTY-TWO

18	22	15
22	17	4
15	22	17
12	7	18
22	11	15

DIRECTIONS: TRACE THE WORDS AND NUMBERS BELOW.

23

Twenty-three

23 23 23

23 23 23

COLOR TWENTY-THREE TRIANGLES

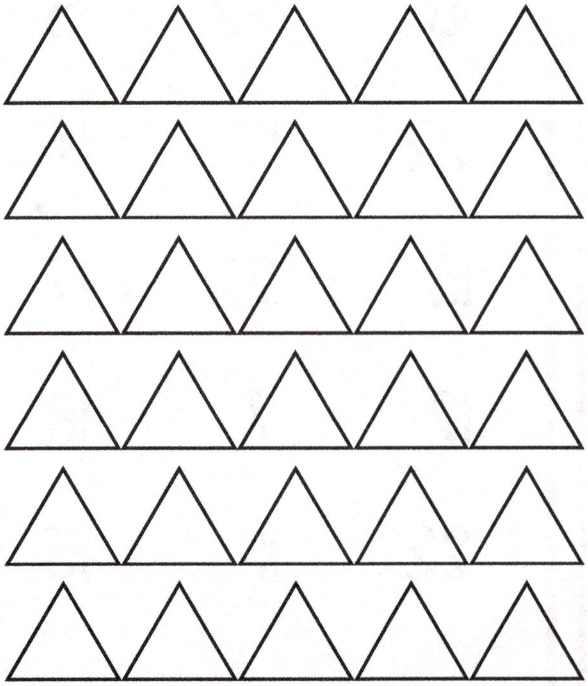

CIRCLE THE TWENTY-THREES

23	14	15
4	23	4
15	18	17
23	7	18
18	23	15

24

DIRECTIONS: TRACE THE WORDS AND NUMBERS BELOW.

Twenty-four

24 24 24

24 24 24

COLOR TWENTY-FOUR STARS

CIRCLE THE TWENTY-FOURS

24	14	15
4	24	4
24	18	17
12	7	18
18	24	15

DIRECTIONS: TRACE THE WORDS AND NUMBERS BELOW.

25

Twenty-five

25 25

25 25 25

COLOR TWENTY-FIVE CIRCLES

CIRCLE THE TWENTY-FIVES

25	14	15
17	25	17
15	16	8
12	25	13
25	10	15

DIRECTIONS: TRACE THE WORDS AND NUMBERS BELOW.

26

Twenty-six

26 26

26 26 26

COLOR TWENTY-SIXS HEARTS

CIRCLE THE TWENTY-SIXS

26	14	15
4	17	4
15	26	17
12	7	26
26	117	15

DIRECTIONS: TRACE THE WORDS AND NUMBERS BELOW.

27

Twenty-seven

27 27

27 27 27

COLOR TWENTY-SEVEN CIRCLES

CIRCLE THE TWENTY-SEVENS

27	14	27
4	17	4
15	27	17
12	7	18
18	27	15

DIRECTIONS: TRACE THE WORDS AND NUMBERS BELOW.

28

Twenty-eight

28 28

28 28 28

COLOR TWENTY-EIGHT STARS

CIRCLE THE TWENTY-EIGHTS

18	14	28
28	17	4
15	28	17
12	7	18
28	17	15

DIRECTIONS: TRACE THE WORDS AND NUMBERS BELOW.

29

Twenty-nine

29 29

29 29 29

COLOR TWENTY-NINE HEXAGONS

CIRCLE THE TWENTY-NINES

18	29	15
4	17	4
29	18	17
12	7	29
18	29	15

DIRECTIONS: TRACE THE WORDS AND NUMBERS BELOW.

30

Thirty

30 30

30 30 30

COLOR THIRTY CIRCLES:

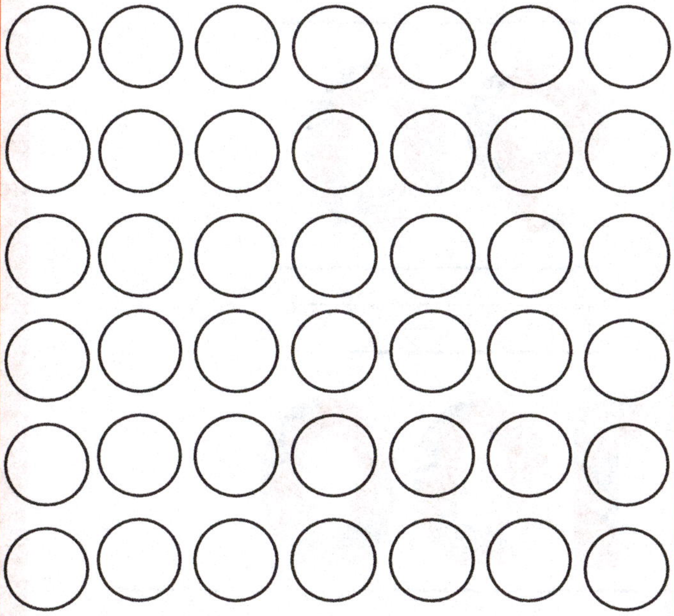

CIRCLE THE THIRTYS

18	30	15
4	17	4
30	18	17
12	30	18
30	117	15

NUMBERS ten by ten

Read and trace the numbers.

10
ten

20
twenty

30
thirty

40
forty

50
fifty

60
sixty

70
seventy

80
eighty

90
ninety

100
one hundred

DIRECTIONS: TRACE THE WORDS AND NUMBERS BELOW.

31

Thirty-one

31 31 31

31 31 31 31

COLOR THIRTY-ONE CIRCLES:

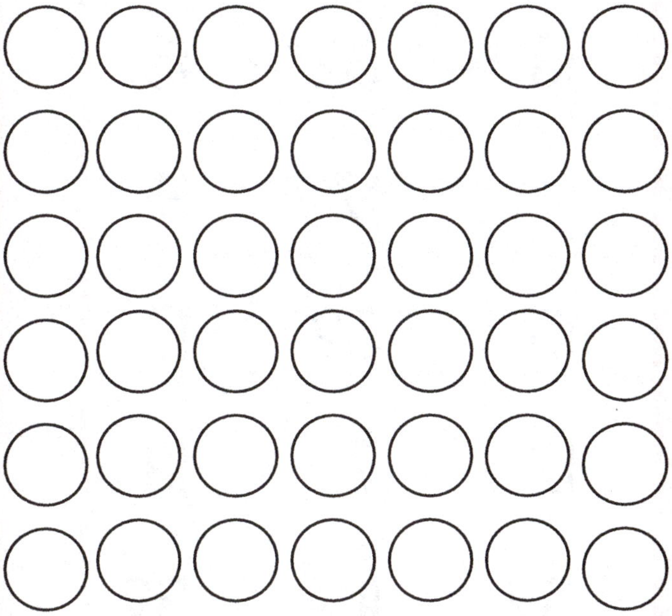

CIRCLE THE THIRTY-ONES

18	31	15
4	17	4
15	31	17
12	7	18
31	11	15

DIRECTIONS: TRACE THE WORDS AND NUMBERS BELOW.

32

thirty-two

32 32

32 32 32

COLOR THIRTY-TWO STARS

CIRCLE THE THIRTY-TWOS

18	32	15
4	17	4
15	32	17
12	7	18
32	11	32

DIRECTIONS: TRACE THE WORDS AND NUMBERS BELOW.

33

Thirty-three

33 33

33 33 33

COLOR THIRTY-THREE STAR:S

CIRCLE THE THIRTY-THREES

18	14	15
4	33	4
15	18	17
33	7	33
18	33	15

DIRECTIONS: TRACE THE WORDS AND NUMBERS BELOW.

34

Thirty-four

34 34

34 34 34

COLOR THIRTY-FOUR STAR:S

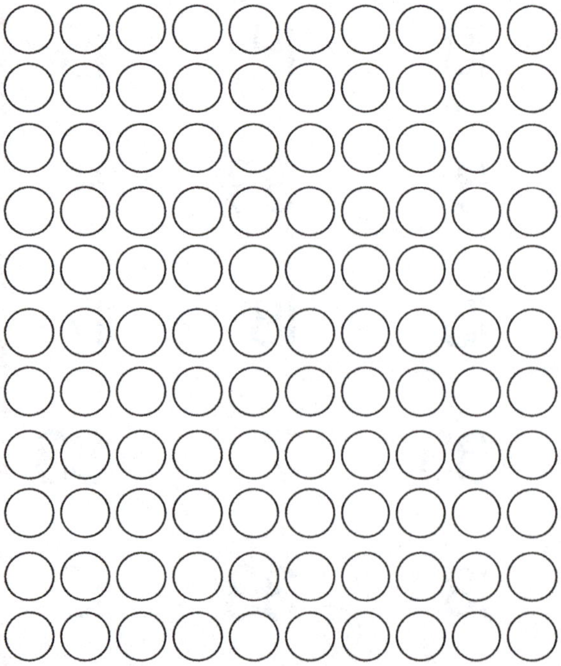

CIRCLE THE THIRTY-FOURS

34	14	15
4	17	34
15	18	17
12	34	18
34	117	15

DIRECTIONS: TRACE THE WORDS AND NUMBERS BELOW.

35

Thirty-five

35 35

35 35 35

COLOR THIRTY-FIVE STAR:S

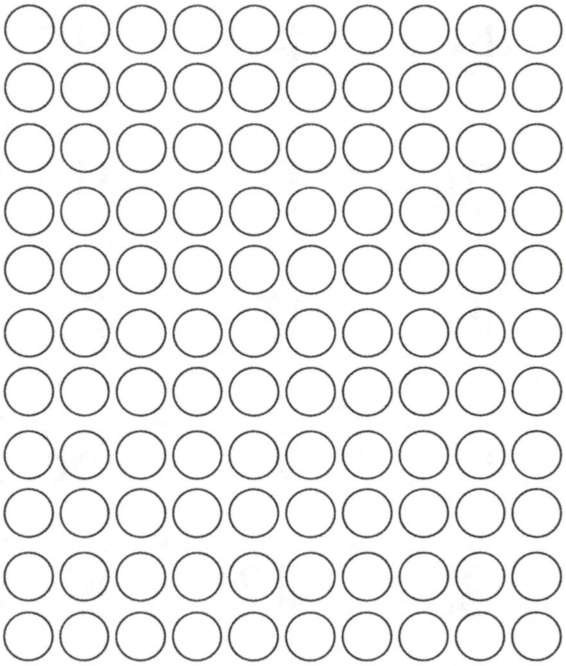

CIRCLE THE THIRTY-FIVES

35	14	15
4	17	4
15	35	17
12	7	18
18	117	35

DIRECTIONS: TRACE THE WORDS AND NUMBERS BELOW.

36

Thirty-six

36 36

36 36 36

COLOR THIRTY-SIX STAR:S

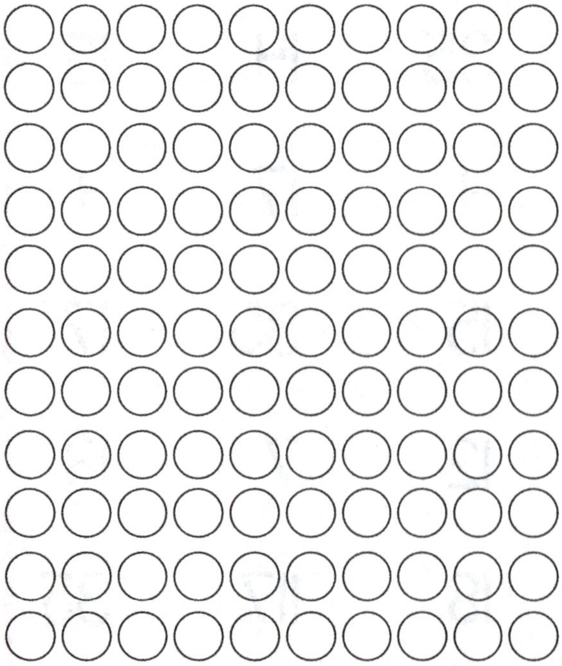

CIRCLE THE THIRTY-SIXS

18	14	36
36	17	4
15	18	17
12	7	36
36	117	15

DIRECTIONS: TRACE THE WORDS AND NUMBERS BELOW.

37

Thirty-seven

37 37

37 37 37

COLOR THIRTY-SEVEN STAR:S

CIRCLE THE THIRTY-SEVENS

37	14	15
4	37	4
15	18	37
12	7	18
18	37	15

DIRECTIONS: TRACE THE WORDS AND NUMBERS BELOW.

38

Thirty-eight

38 38

38 38 38

COLOR THIRTY-EIGHT STAR:S

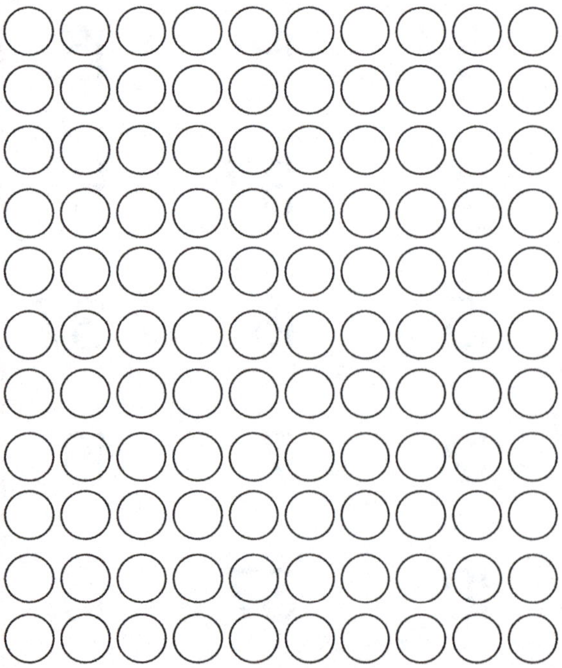

CIRCLE THE THIRTY-EIGHTS

38	14	15
4	38	4
15	18	17
12	38	18
38	117	15

DIRECTIONS: TRACE THE WORDS AND NUMBERS BELOW.

39

Thirty-nine

39 39

39 39 39

COLOR THIRTY-NINE CIRCLES

CIRCLE THE THIRTY-NINES

18	39	15
4	17	4
15	39	17
12	7	18
18	39	15

DIRECTIONS: TRACE THE WORDS AND NUMBERS BELOW.

40

Forty

40 40

40 40 40

COLOR FORTY CIRCLES

CIRCLE THE FORTYS

40	14	15
4	17	4
15	40	17
40	7	18
18	40	15

DIRECTIONS: TRACE THE WORDS AND NUMBERS BELOW.

41

Forty-one

41 41 41

41 41 41 41

COLOR FORTY CIRCLES

CIRCLE THE FORTY-ONES

41	14	15
4	41	4
15	18	17
41	7	41
41	11	15

DIRECTIONS: TRACE THE WORDS AND NUMBERS BELOW.

42

Forty-two

42 42

42 42 42

COLOR FORTY-TWO CIRCLES

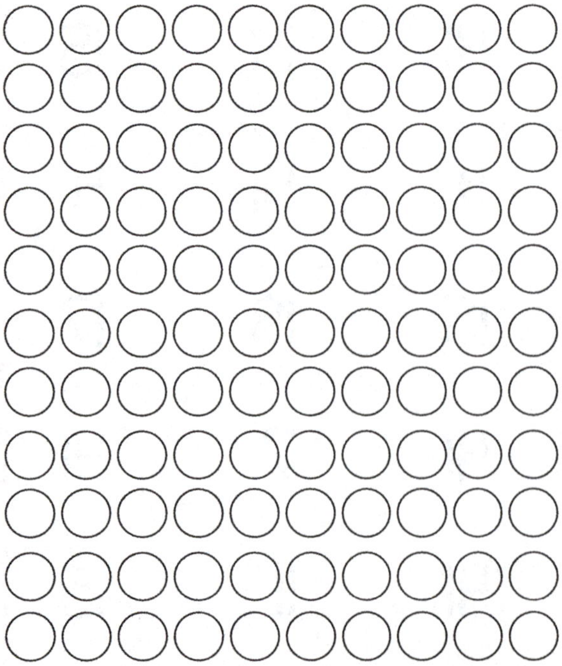

CIRCLE THE FORTY-TWOS

42	14	15
4	42	4
42	18	17
12	7	18
18	42	15

DIRECTIONS: TRACE THE WORDS AND NUMBERS BELOW.

43

Forty-three

43 43

43 43 43

COLOR FORTY-THREE CIRCLES

CIRCLE THE FORTY-THREES

43	14	15
4	43	4
15	18	17
43	7	18
18	117	43

DIRECTIONS: TRACE THE WORDS AND NUMBERS BELOW.

44

Forty-four

44 44

44 44 44

COLOR FORTY-FOUR CIRCLES

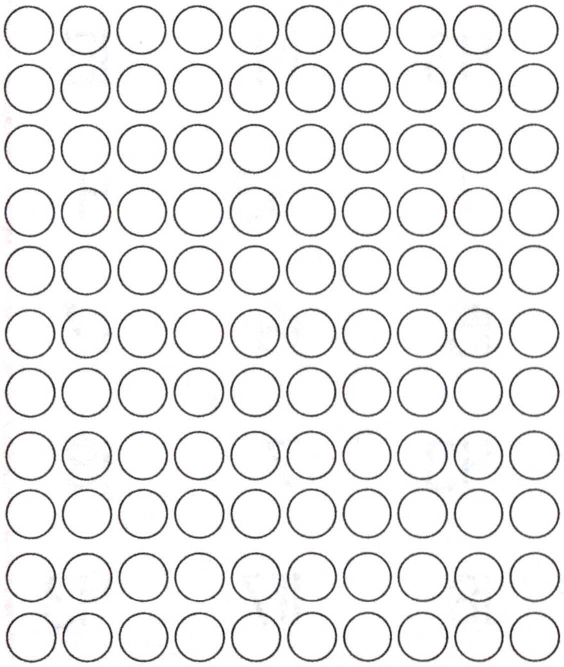

CIRCLE THE FORTY-FOURS

44	14	15
4	44	4
15	18	17
12	44	18
44	117	15

DIRECTIONS: TRACE THE WORDS AND NUMBERS BELOW.

45

Forty-five

45 45

45 45 45

COLOR FORTY-FIVE CIRCLES

CIRCLE THE FORTY-FIVES

18	45	15
4	40	4
12	18	17
12	7	18
45	117	45

DIRECTIONS: TRACE THE WORDS AND NUMBERS BELOW.

46

Forty-six

46 46

46 46 46

COLOR FORTY-SIX CIRCLES

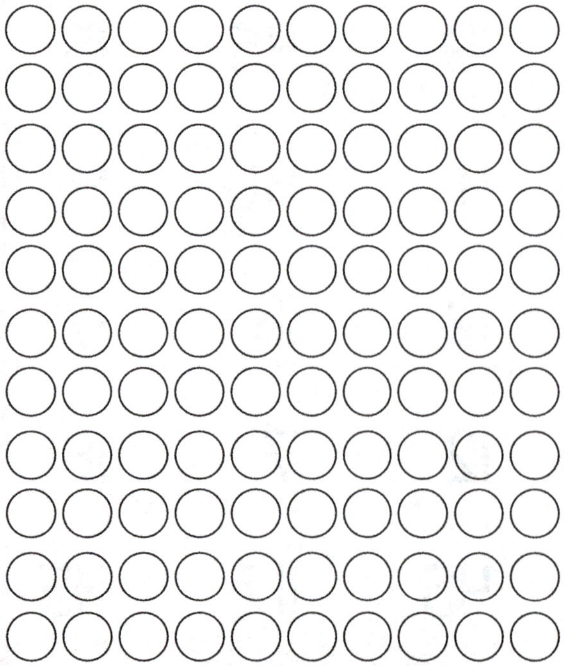

CIRCLE THE FORTY-SIXS

18	14	15
4	46	4
15	18	17
12	46	18
46	117	15

DIRECTIONS: TRACE THE WORDS AND NUMBERS BELOW.

47

Forty-seven

47 47

47 47 47

COLOR FORTY-SEVEN CIRCLES

CIRCLE THE FORTY-SEVENS

47	14	15
4	17	4
15	47	17
12	7	18
18	47	15

DIRECTIONS: TRACE THE WORDS AND NUMBERS BELOW.

48

Forty-eight

48 48

48 48 48

COLOR FORTY-EIGHT CIRCLES

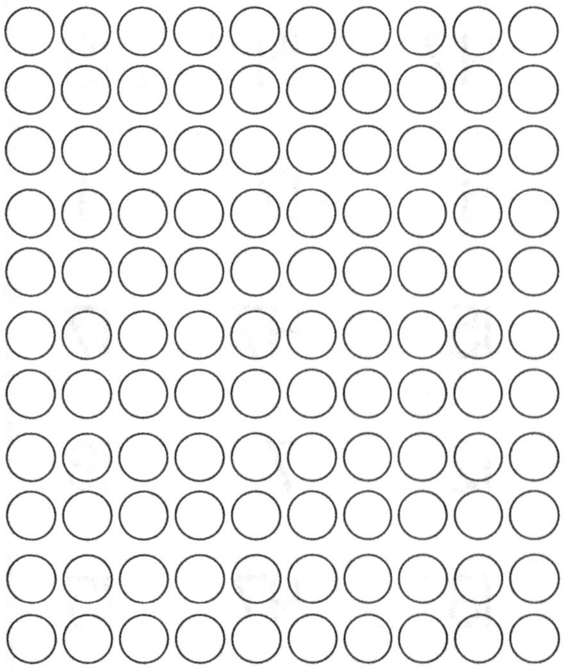

CIRCLE THE FORTY-EIGHTS

48	14	15
4	48	4
15	18	17
12	48	18
48	117	15

DIRECTIONS: TRACE THE WORDS AND NUMBERS BELOW.

49

Forty-nine

49 49

49 49 49

COLOR FORTY-NINE CIRCLES

CIRCLE THE FORTY-NINES

18	14	15
4	49	4
41	18	17
49	7	18
18	11	49

DIRECTIONS: TRACE THE WORDS AND NUMBERS BELOW.

50

Fifty

50 50

50 50 50

COLOR FIFTY CIRCLES

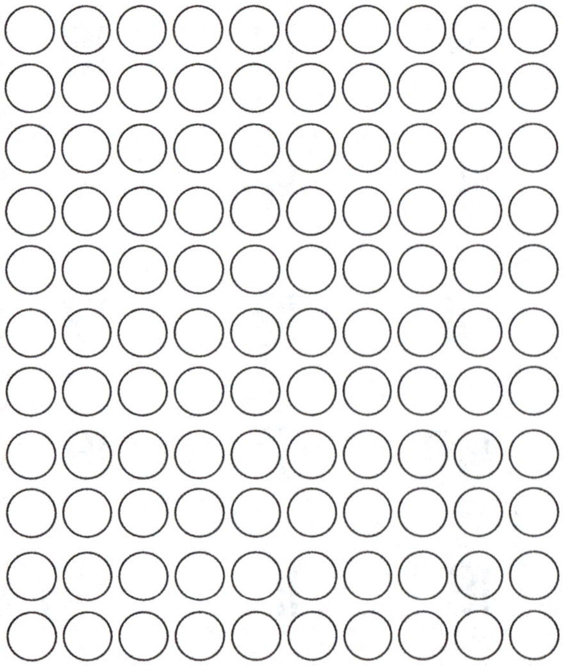

CIRCLE THE FIFTYS

18	14	15
4	50	4
15	18	17
12	50	18
50	117	15

DIRECTIONS: TRACE THE WORDS AND NUMBERS BELOW.

51

Fifty-one

51 51 51

51 51 51 51

COLOR FIFTY-ONE CIRCLES

CIRCLE THE FIFTY-ONES

51	14	15
4	17	4
15	51	17
12	7	51
51	117	15

DIRECTIONS: TRACE THE WORDS AND NUMBERS BELOW.

52

Fifty-two

52 52

52 52 52

COLOR FIFTY-TWO CIRCLES

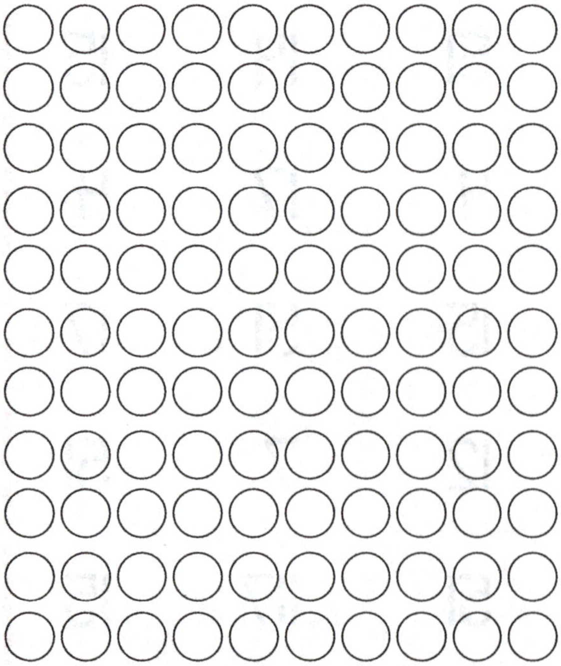

CIRCLE THE FIFTY-TWOS

52	14	15
4	17	4
15	52	17
12	7	18
18	52	15

DIRECTIONS: TRACE THE WORDS AND NUMBERS BELOW.

53

Fifty-three

53 53

53 53 53

COLOR FIFTY-THREE CIRCLES

CIRCLE THE FIFTY-THREES

53	14	53
4	17	4
15	18	17
12	53	18
18	53	15

DIRECTIONS: TRACE THE WORDS AND NUMBERS BELOW.

54

fifty-four

54 54

54 54 54

COLOR FIFTY-FOUR CIRCLES

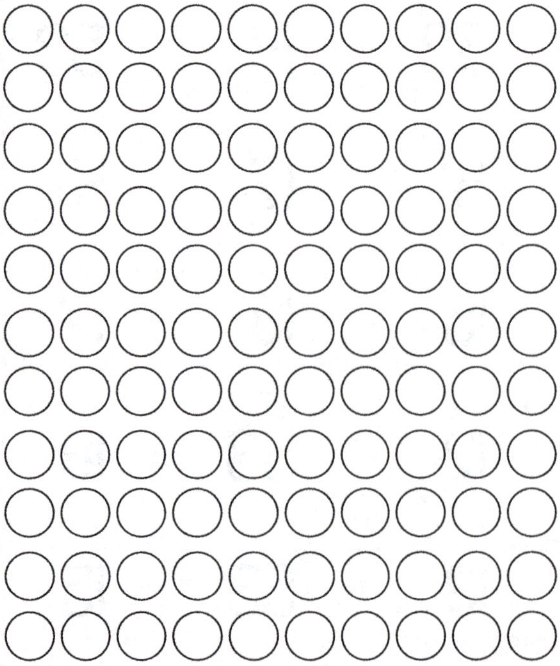

CIRCLE THE FIFTY-FOURS

54	14	15
4	17	4
15	54	17
12	7	18
18	54	15

DIRECTIONS: TRACE THE WORDS AND NUMBERS BELOW.

55

Fifty-five

55 55

55 55 55

COLOR FIFTY-FIVE CIRCLES

CIRCLE THE FIFTY-FIVES

55	14	15
4	17	4
15	55	17
12	7	18
18	55	15

DIRECTIONS: TRACE THE WORDS AND NUMBERS BELOW.

56

Fifty-six

56 56

56 56 56

COLOR FIFTY-SIX CIRCLES

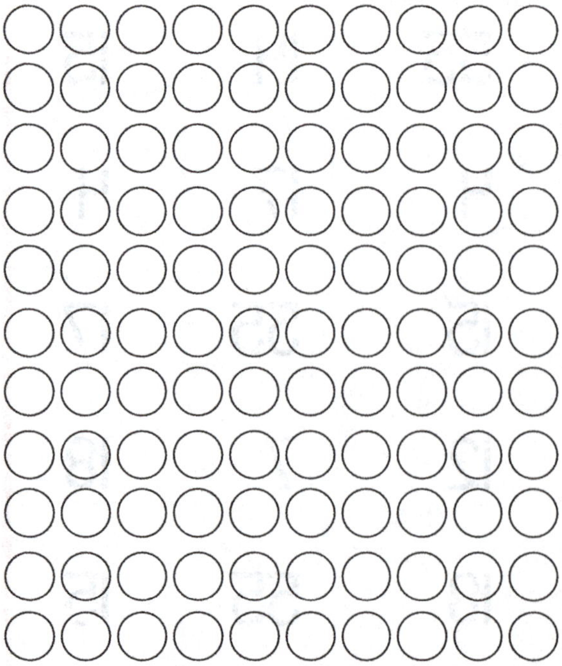

CIRCLE THE FIFTY-SIXS

56	14	56
4	17	4
15	56	17
12	7	18
56	117	15

DIRECTIONS: TRACE THE WORDS AND NUMBERS BELOW.

57

Fifty-seven

57 57

57 57 57

COLOR FIFTY-SEVEN CIRCLES

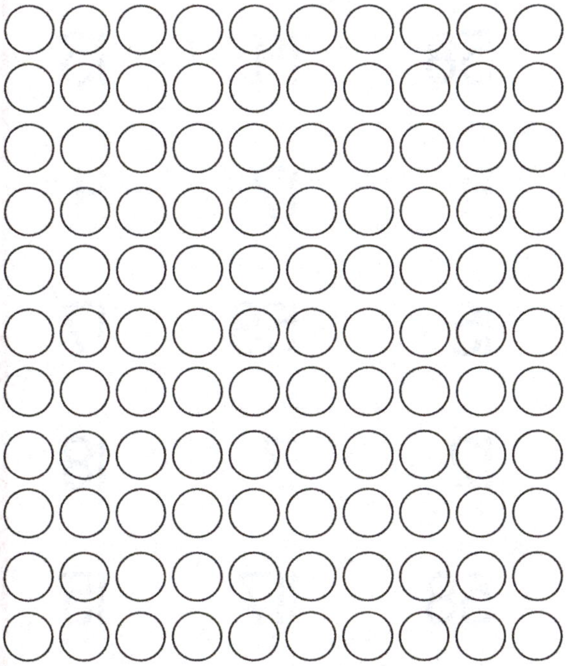

CIRCLE THE FIFTY-SEVENS

57	14	15
4	17	4
15	57	17
12	7	18
18	57	15

DIRECTIONS: TRACE THE WORDS AND NUMBERS BELOW.

58

Fifty-eight

58 58

58 58 58

COLOR FIFTY-EIGHT CIRCLES

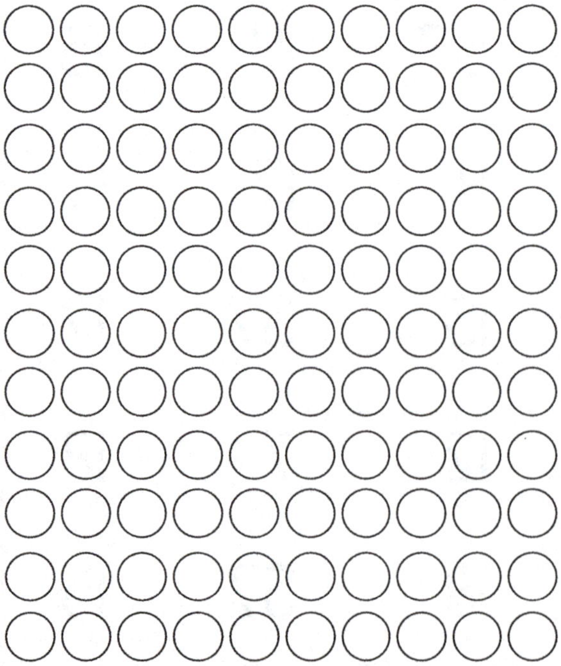

CIRCLE THE FIFTY-EIGHTS

58	14	15
4	17	4
15	58	17
12	7	18
58	117	15

DIRECTIONS: TRACE THE WORDS AND NUMBERS BELOW.

59

Fifty-nine

59 59

59 59 59

COLOR FIFTY-NINE CIRCLES

CIRCLE THE FIFTY-NINES

18	14	15
4	59	4
15	18	17
12	59	18
59	117	15

DIRECTIONS: TRACE THE WORDS AND NUMBERS BELOW.

60

Sixty

60 60

60 60 60

COLOR SIXTY CIRCLES

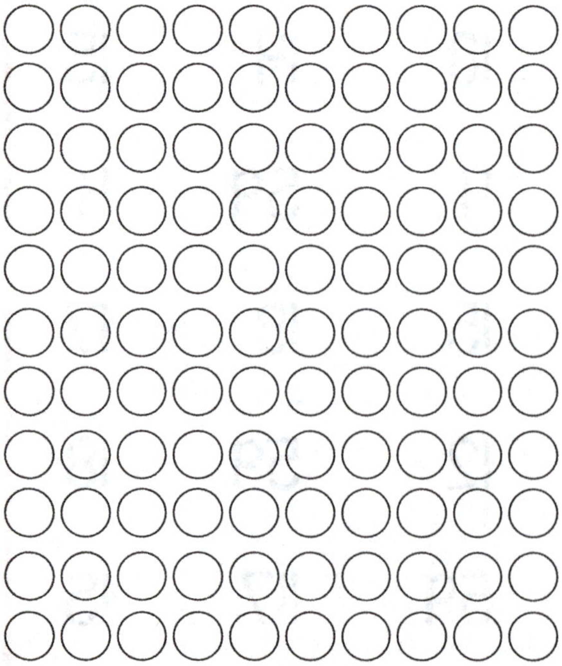

CIRCLE THE SIXTYS

60	14	15
4	17	4
15	60	17
60	7	18
18	117	15

DIRECTIONS: TRACE THE WORDS AND NUMBERS BELOW.

61

Sixty-one

61 61 61

61 61 61

COLOR SIXTY-ONE STAR:

CIRCLE THE SIXTY-ONES

61	14	15
4	17	4
15	61	17
12	7	18
18	61	15

DIRECTIONS: TRACE THE WORDS AND NUMBERS BELOW.

62

Sixty-two

62 62

62 62 62

COLOR SIXTY-TWO STAR:

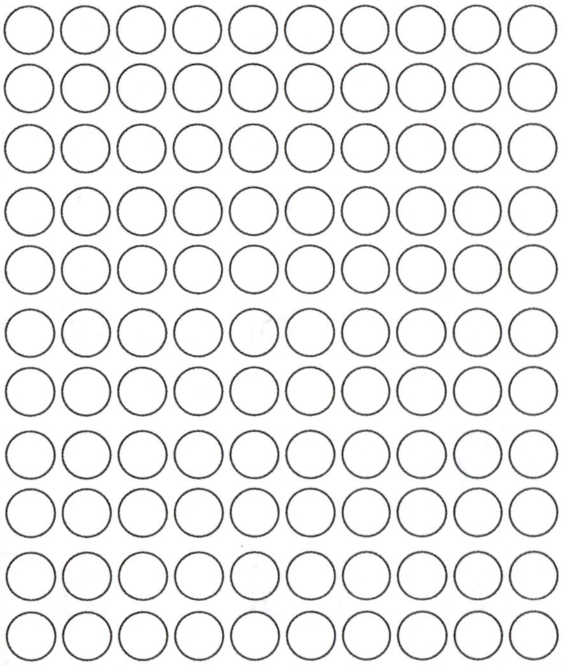

CIRCLE THE SIXTY-TWOS

62	14	62
4	17	4
62	18	17
12	7	18
18	62	15

63

DIRECTIONS: TRACE THE WORDS AND NUMBERS BELOW.

Sixty-three

63 63

63 63 63

COLOR SIXTY-THREE STAR:

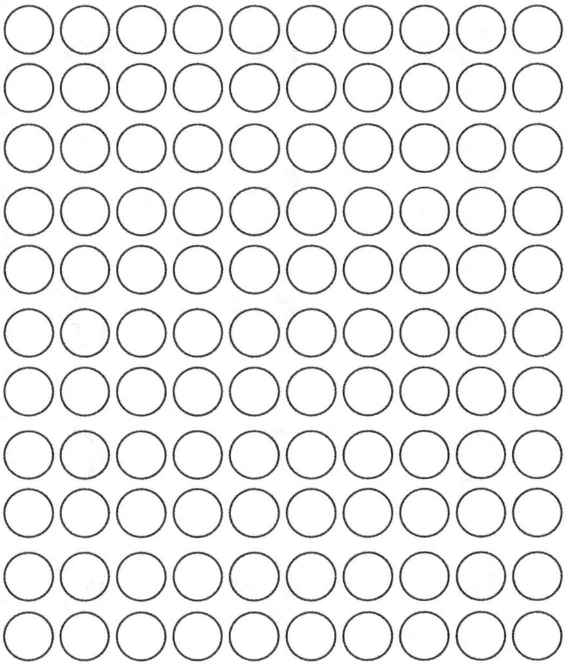

CIRCLE THE SIXTY-THREES

63	14	15
4	17	4
15	63	17
12	7	18
63	117	63

DIRECTIONS: TRACE THE WORDS AND NUMBERS BELOW.

64

Sixty-four

64 64

64 64 64

COLOR SIXTY-FOUR STAR:

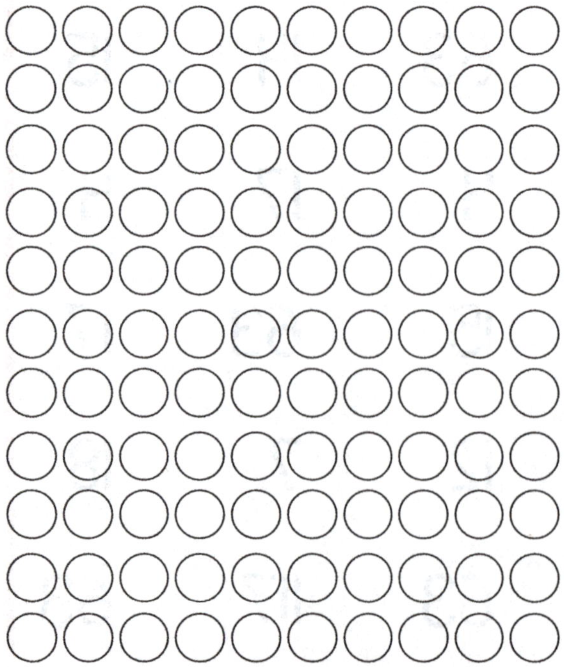

CIRCLE THE SIXTY-FOURS

64	14	15
4	17	4
15	64	17
12	7	18
18	64	15

DIRECTIONS: TRACE THE WORDS AND NUMBERS BELOW.

65

Sixty-five

65 65

65 65 65

COLOR SIXTY-FIVE STAR:

CIRCLE THE SIXTY-FIVES

18	14	15
4	65	4
15	18	17
65	7	18
18	117	65

DIRECTIONS: TRACE THE WORDS AND NUMBERS BELOW.

66

Sixty-six

66 66

66 66 66

COLOR SIXTY-SIX STAR:

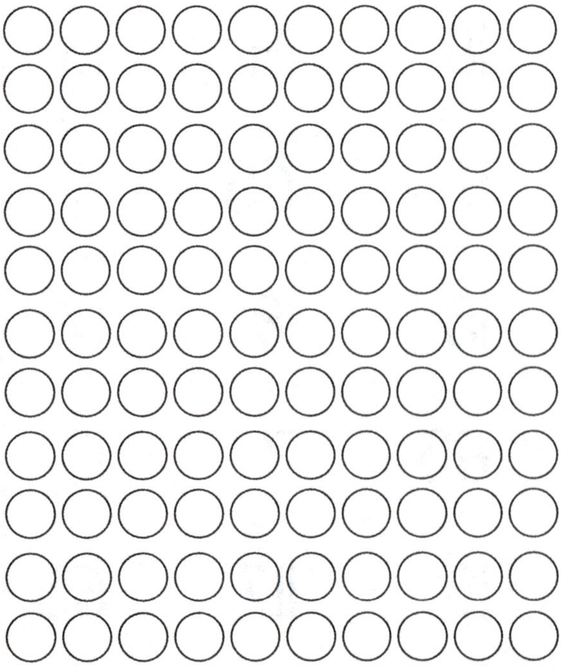

CIRCLE THE SIXTY-SIXS

18	14	15
4	66	4
66	18	17
12	7	18
18	66	15

DIRECTIONS: TRACE THE WORDS AND NUMBERS BELOW.

67

Sixty-seven

67 67

67 67 67

COLOR SIXTY-SEVEN CIRCLES

CIRCLE THE SIXTY-SEVENS

67	14	67
4	17	4
15	67	17
12	7	18
67	117	15

DIRECTIONS: TRACE THE WORDS AND NUMBERS BELOW.

68

Sixty-eight

68 68

68 68 68

COLOR SIXTY-EIGHT CIRCLES

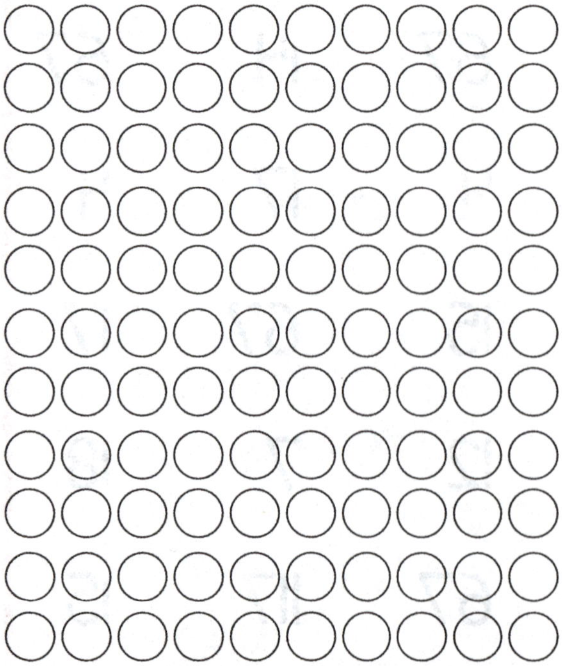

CIRCLE THE SIXTY-EIGHTS

68	14	15
4	17	4
15	68	17
12	7	18
68	117	15

DIRECTIONS: TRACE THE WORDS AND NUMBERS BELOW.

69

Sixty-nine

69 69

69 69 69

COLOR SIXTY-NINE CIRCLES

CIRCLE THE SIXTY-NINES

69	14	15
4	69	4
15	18	17
12	69	18
69	117	15

DIRECTIONS: TRACE THE WORDS AND NUMBERS BELOW.

70

Seventy

70 70

70 70 70

COLOR SEVENTY CIRCLES

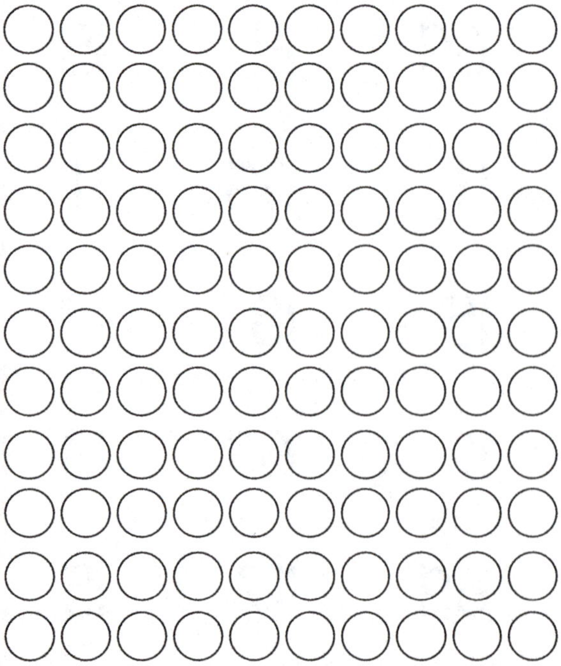

CIRCLE THE SEVENTYS

18	14	70
4	17	4
70	18	17
12	7	18
18	70	15

DIRECTIONS: TRACE THE WORDS AND NUMBERS BELOW.

71

Seventy-one

71 71 71

71 71 71 71

COLOR SEVENTY-ONE CIRCLES

CIRCLE THE SEVENTY-ONES

71	14	15
4	71	4
15	18	17
12	71	18
71	117	15

DIRECTIONS: TRACE THE WORDS AND NUMBERS BELOW.

72

Seventy-two

72 72

72 72 72

COLOR SEVENTY-TWO CIRCLES

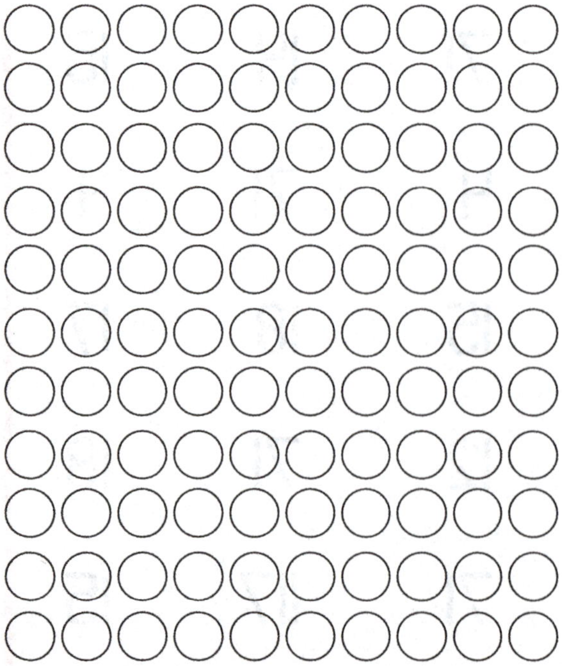

CIRCLE THE SEVENTY-TWOS

72	14	15
4	17	4
15	72	17
12	7	18
72	117	72

DIRECTIONS: TRACE THE WORDS AND NUMBERS BELOW.

73

Seventy-three

73 73

73 73 73

COLOR SEVENTY-THREE CIRCLES

CIRCLE THE SEVENTY-THREES

18	73	15
4	17	4
15	73	17
12	7	18
18	73	15

DIRECTIONS: TRACE THE WORDS AND NUMBERS BELOW.

74

Seventy-four

74 74

74 74 74

COLOR SEVENTY-FOUR CIRCLES

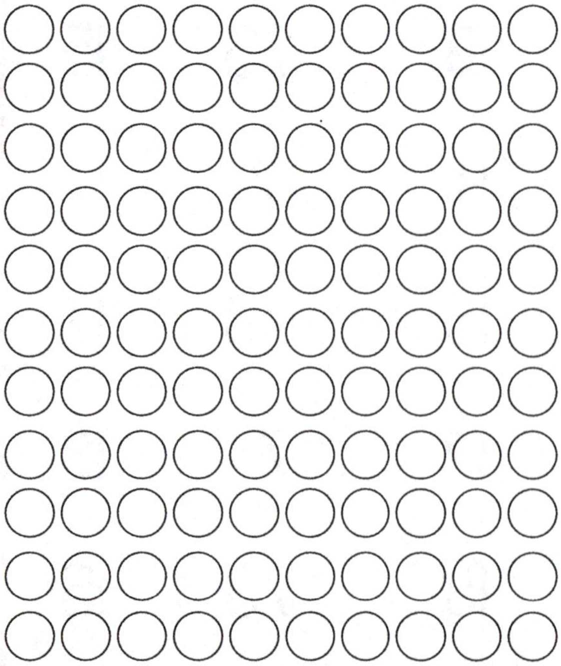

CIRCLE THE SEVENTY-FOURS

74	14	15
4	17	4
15	74	17
12	7	18
74	117	74

DIRECTIONS: TRACE THE WORDS AND NUMBERS BELOW.

75

Seventy-five

75 75

75 75 75

COLOR SEVENTY-FIVE CIRCLES

CIRCLE THE SEVENTY-FIVES

75	14	15
4	74	4
15	18	17
75	7	18
18	117	75

76

DIRECTIONS: TRACE THE WORDS AND NUMBERS BELOW.

Seventy-six

76 76

76 76 76

COLOR SEVENTY-SIX CIRCLES

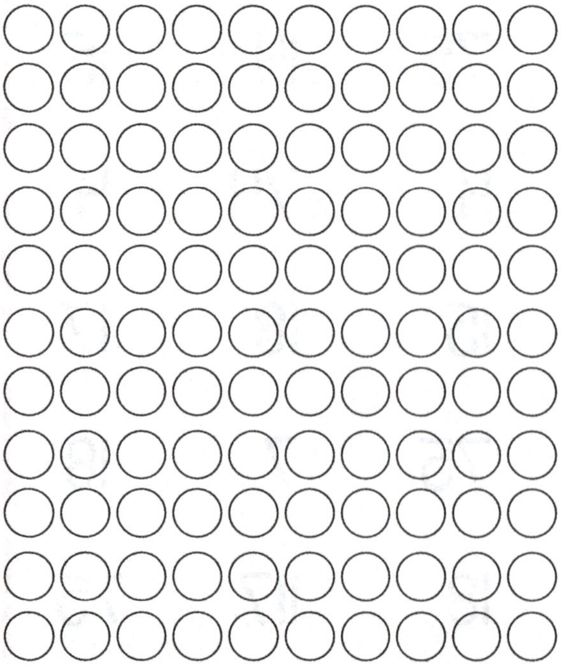

CIRCLE THE SEVENTY-SIXS

76	14	15
4	76	4
15	18	17
12	76	18
76	117	15

DIRECTIONS: TRACE THE WORDS AND NUMBERS BELOW.

77

Seventy-seven

77 77

77 77 77

COLOR SEVENTY-SEVEN CIRCLES

CIRCLE THE SEVENTY-SEVENS

18	14	15
4	77	4
15	18	17
12	77	18
77	117	15

DIRECTIONS: TRACE THE WORDS AND NUMBERS BELOW.

78

Seventy-eight

78 78

78 78 78

COLOR SEVENTY-EIGHT CIRCLES

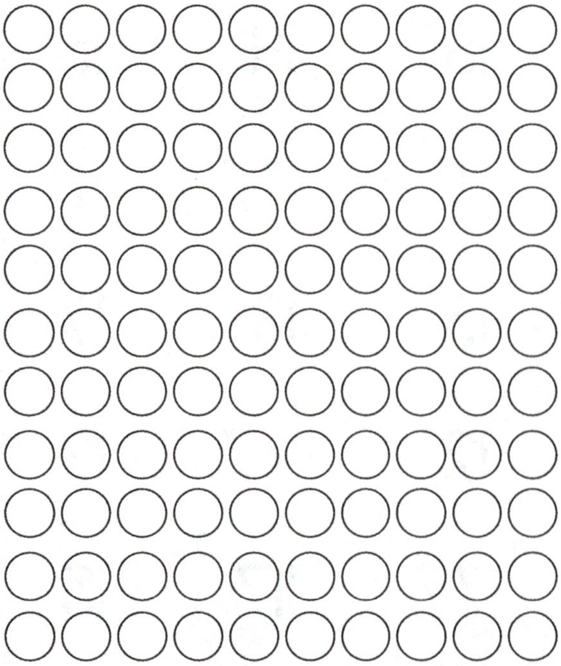

CIRCLE THE SEVENTY-EIGHTS

18	14	15
4	78	4
15	18	17
12	78	18
78	117	15

DIRECTIONS: TRACE THE WORDS AND NUMBERS BELOW.

79

Seventy-nine

79 79

79 79 79

COLOR SEVENTY-NINE CIRCLES

CIRCLE THE SEVENTY-NINES

79	14	15
4	17	4
15	79	17
12	7	18
79	117	15

DIRECTIONS: TRACE THE WORDS AND NUMBERS BELOW.

80

Eighty

80 80

80 80 80

COLOR EIGHTY CIRCLES

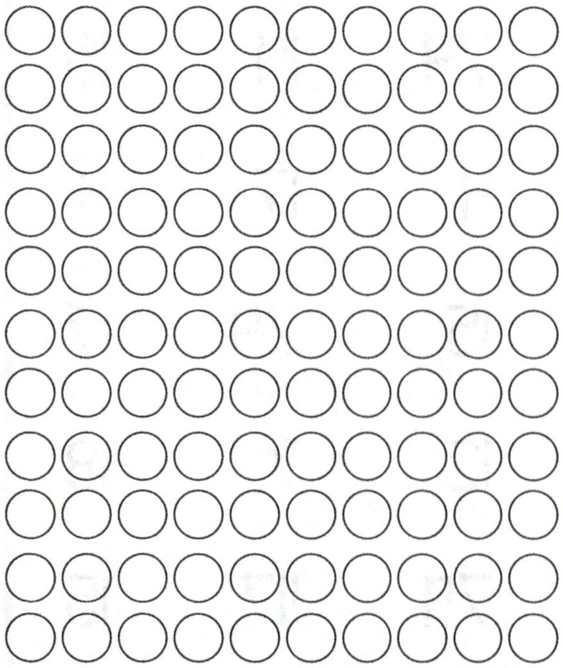

CIRCLE THE EIGHTYS

80	14	15
4	17	4
15	80	17
12	7	18
80	117	15

DIRECTIONS: TRACE THE WORDS AND NUMBERS BELOW.

81

Eighty-one

81 81 81

81 81 81 81

COLOR EIGHTY-ONE CIRCLES

CIRCLE THE EIGHTY-ONES

81	14	15
4	81	4
15	18	17
81	7	18
18	117	81

DIRECTIONS: TRACE THE WORDS AND NUMBERS BELOW.

82

Eighty-two

82 82

82 82 82

COLOR EIGHTY-TWO CIRCLES

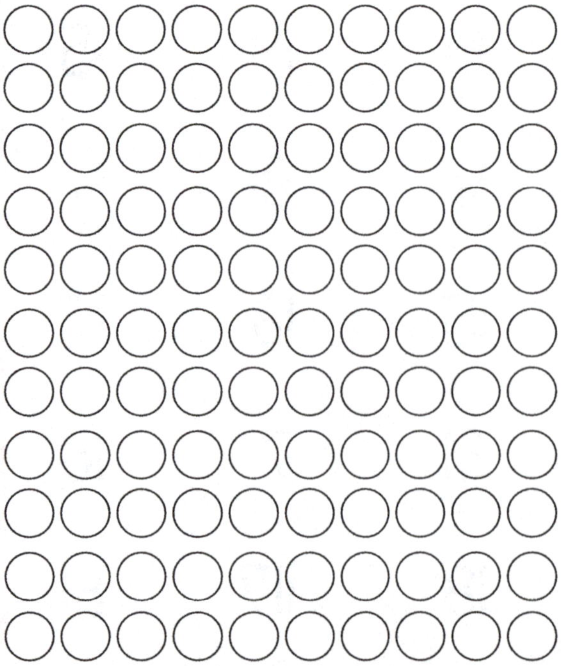

CIRCLE THE EIGHTY-TWOS

82	14	15
4	82	4
15	18	17
12	82	18
82	117	15

DIRECTIONS: TRACE THE WORDS AND NUMBERS BELOW.

83

Eighty-three

83 83

83 83 83

COLOR EIGHTY-THREE CIRCLES

CIRCLE THE EIGHTY-THREES

83	14	15
4	17	4
15	83	17
12	7	18
83	117	15

DIRECTIONS: TRACE THE WORDS AND NUMBERS BELOW.

84

Eighty-four

84 84

84 84 84

COLOR EIGHTY-FOUR CIRCLES

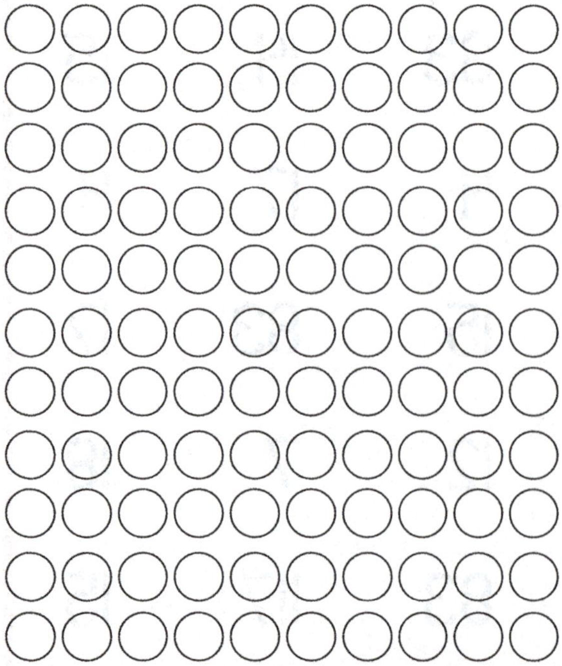

CIRCLE THE EIGHTY-FOURS

18	84	15
4	17	4
15	84	17
12	7	18
84	117	15

85

DIRECTIONS: TRACE THE WORDS AND NUMBERS BELOW.

Eighty-five

85 85

85 85 85

COLOR EIGHTY-FIVE CIRCLES

CIRCLE THE EIGHTY-FIVES

85	14	15
4	17	4
15	85	17
12	7	18
18	85	15

DIRECTIONS: TRACE THE WORDS AND NUMBERS BELOW.

86

Eighty-six

8 6 8 6

8 6 8 6 8 6

COLOR EIGHTY-SIX CIRCLES

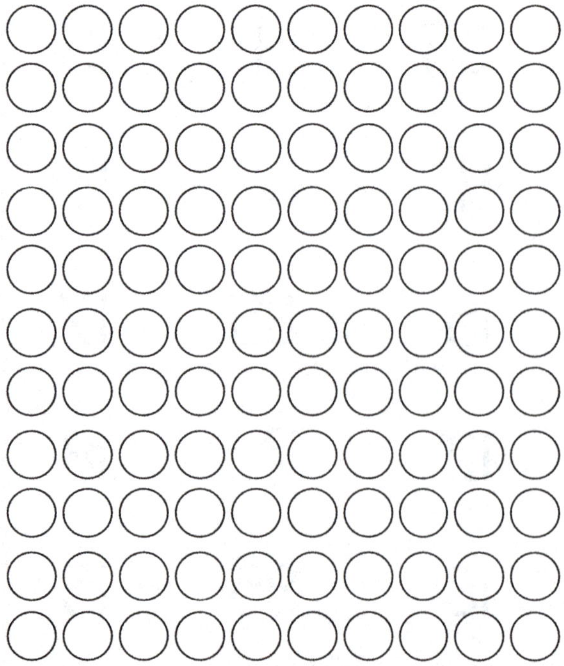

CIRCLE THE EIGHTY-SIXS

18	86	15
4	17	4
15	18	17
12	86	18
18	117	86

DIRECTIONS: TRACE THE WORDS AND NUMBERS BELOW.

87

Eighty-seven

87 87

87 87 87

COLOR EIGHTY-SEVEN CIRCLES

CIRCLE THE EIGHTY-SEVENS

87	14	15
4	17	4
15	87	17
12	7	18
18	87	15

88

DIRECTIONS: TRACE THE WORDS AND NUMBERS BELOW.

eighty-eight

88 88

88 88 88

COLOR EIGHTY-EIGHT CIRCLES

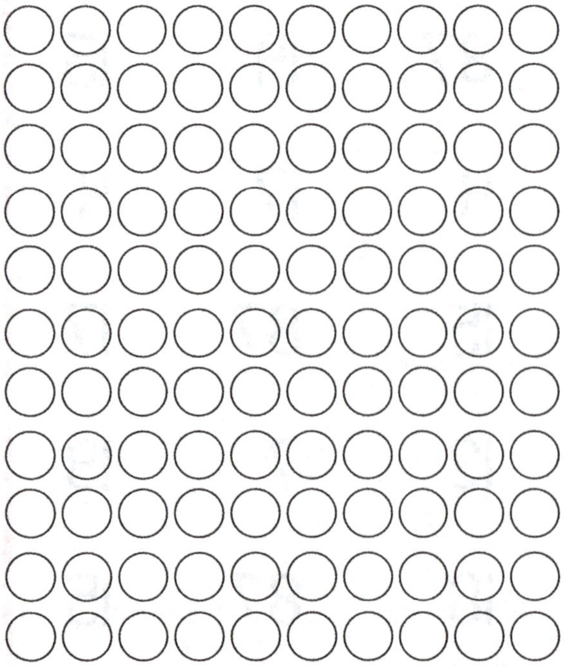

CIRCLE THE EIGHTY-EIGHTS

88	14	15
4	17	4
15	88	17
12	7	18
18	88	15

DIRECTIONS: TRACE THE WORDS AND NUMBERS BELOW.

89

Eighty-nine

89 89 89

89 89 89 89

COLOR EIGHTY-NINE CIRCLES

CIRCLE THE EIGHTY-NINES

89	14	15
4	17	4
15	89	17
12	7	18
89	117	15

DIRECTIONS: TRACE THE WORDS AND NUMBERS BELOW.

90

Ninety

90 90

90 90 90

COLOR NINETY CIRCLES

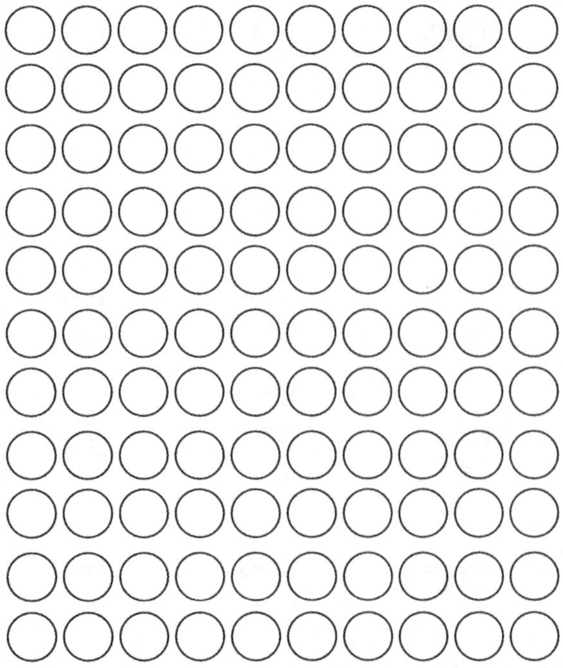

CIRCLE THE NINETYS

90	14	15
4	17	4
15	90	17
12	7	18
18	90	15

DIRECTIONS: TRACE THE WORDS AND NUMBERS BELOW.

91

Ninety-one

91 91 91

91 91 91 91

COLOR NINETY-ONE CIRCLES

CIRCLE THE NINETY-ONES

91	14	15
4	17	4
15	91	17
12	7	18
18	91	15

92

DIRECTIONS: TRACE THE WORDS AND NUMBERS BELOW.

Ninety-two

92 92

92 92 92

COLOR NINETY-TWO CIRCLES

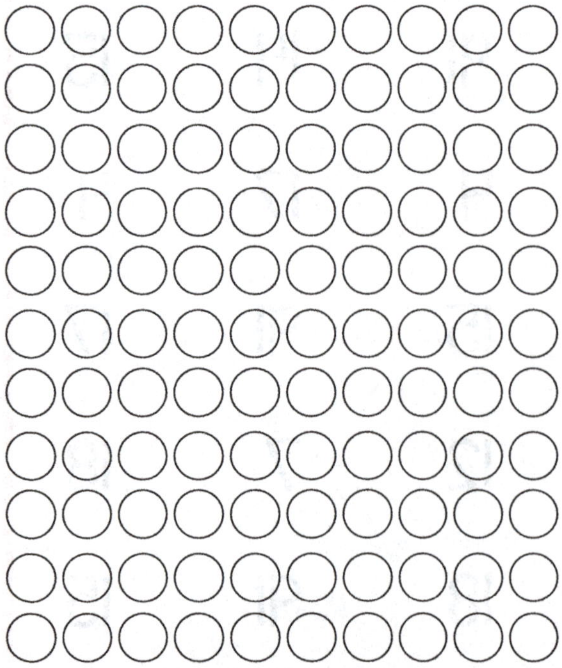

CIRCLE THE NINETY-TWOS

92	14	15
4	92	4
15	18	17
12	92	18
92	117	15

DIRECTIONS: TRACE THE WORDS AND NUMBERS BELOW.

93

Ninety-three

93 93

93 93 93

COLOR NINETY-THREE CIRCLES

CIRCLE THE NINETY-THREES

18	14	15
4	17	4
15	18	17
12	7	18
18	117	15

DIRECTIONS: TRACE THE WORDS AND NUMBERS BELOW.

94

Ninety-four

94 94

94 94 94

COLOR NINETY-FOUR CIRCLES

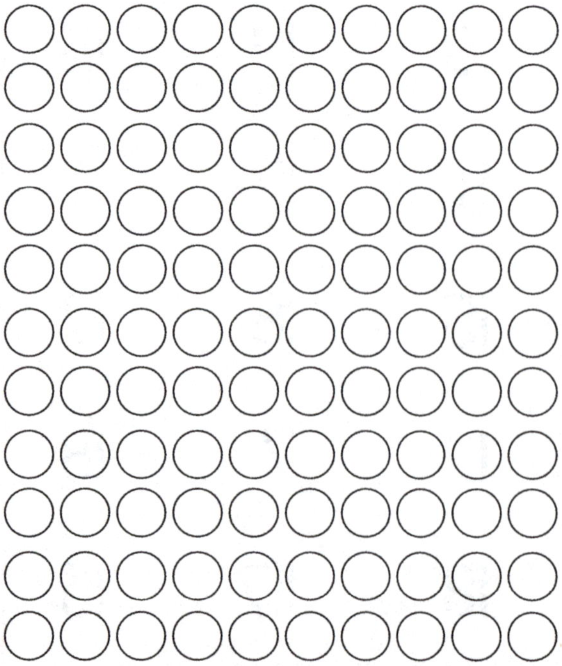

CIRCLE THE NINETY-FOURS

94	14	15
4	17	4
15	94	17
12	7	18
94	117	94

DIRECTIONS: TRACE THE WORDS AND NUMBERS BELOW.

95

Ninety-five

95 95

95 95 95

COLOR NINETY-FIVE CIRCLES

CIRCLE THE NINETY-FIVES

95	14	15
4	17	4
15	95	17
12	7	95
18	95	15

DIRECTIONS: TRACE THE WORDS AND NUMBERS BELOW.

96

Ninety-six

96 96

96 96 96

COLOR NINETY-SIX CIRCLES

CIRCLE THE NINETY-SIXS

18	96	15
4	17	4
15	96	17
12	7	18
96	117	15

DIRECTIONS: TRACE THE WORDS AND NUMBERS BELOW.

97

Ninety-seven

97 97

97 97 97

COLOR NINETY-SEVEN CIRCLES

CIRCLE THE NINETY-SEVENS

18	14	15
4	97	4
15	18	17
12	97	18
97	117	15

DIRECTIONS: TRACE THE WORDS AND NUMBERS BELOW.

98

Ninety-eight

98 98

98 98 98

COLOR NINETY-EIGHT CIRCLES

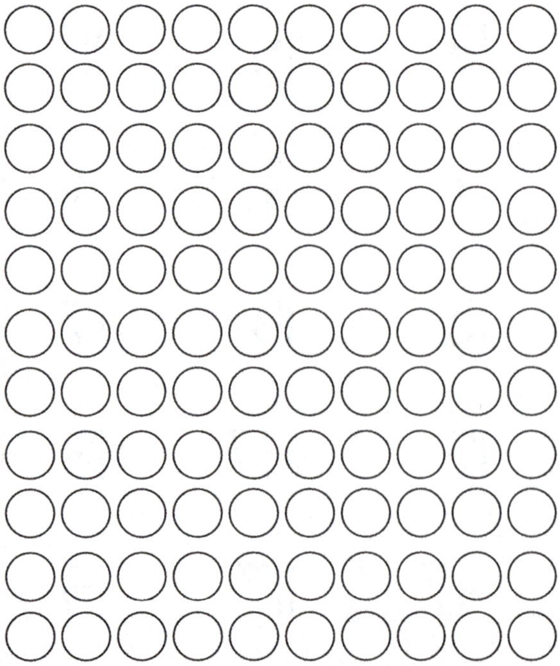

CIRCLE THE NINETY-EIGHTS

98	14	15
4	17	4
15	98	17
12	7	18
98	117	15

DIRECTIONS: TRACE THE WORDS AND NUMBERS BELOW.

99

Ninety-nine

99 99

99 99 99

COLOR NINETY-NINE CIRCLES

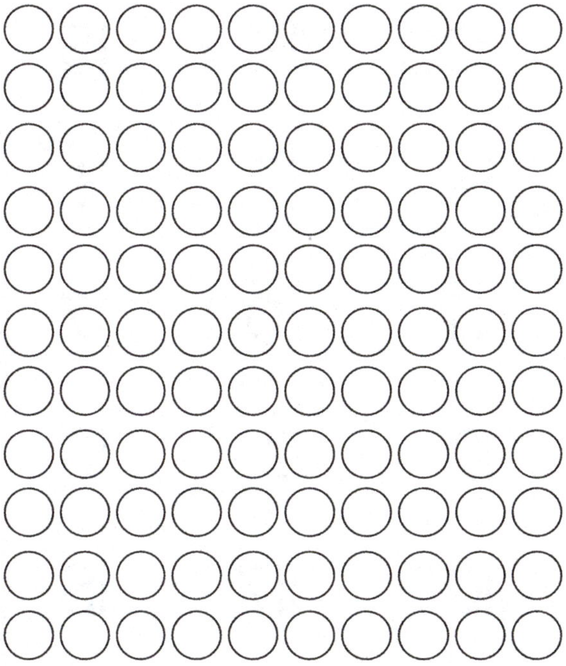

CIRCLE THE NINETY-NINES

99	14	15
4	17	4
15	99	17
12	7	18
18	99	15

DIRECTIONS: TRACE THE WORDS AND NUMBERS BELOW.

100

One Hundred

100 100

100 100 100

COLOR ONE HUNDRED CIRCLES

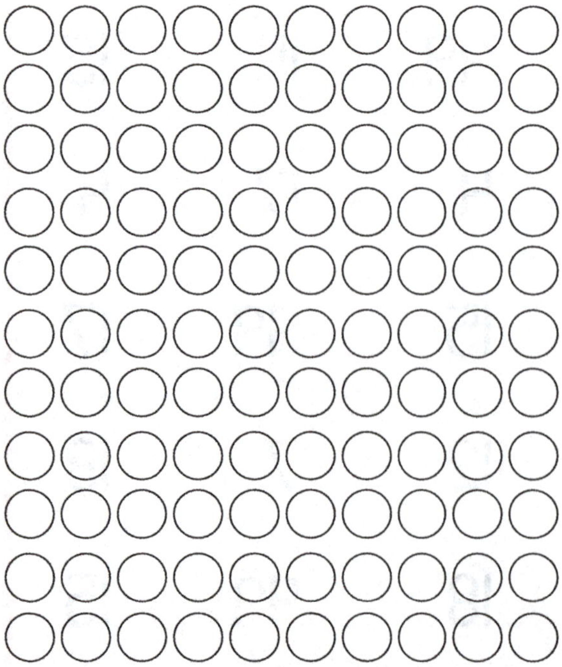

CIRCLE THE ONE HUNDREDS

100	14	15
4	17	4
15	100	17
12	7	18
100	117	15

Made in the USA
Monee, IL
02 November 2026

67381913R00057